Panzerkampfwagen V Panther

Horst Scheibert

A Panther *Ausf.* A. It stands today in the outdoor museum of the Bundeswehr's *Kampftruppenschule* 2.

SCHIFFER MILITARY HISTORY
West Chester, PA

Sources:
Bundesarchiv Koblenz
Podzun-Pallas-Verlag Archives
Scheibert Archives
Generalmajor (Rtd.) Dr. F. Bäke
Kampftruppenschule 2 of the Bundeswehr
Arms and Armour Press

Translated from the German by David Johnston.

Library of Congress Catalog Number: 91-60861.

Printed in the United States of America.
ISBN: 0-88740-314-X

This title was originally published under the title,
Panther - Panzerkampfwagen V,
by Podzun-Pallas-Verlag GmbH, 6360 Friedberg 3 (Dorheim).
ISBN: 307909-0062-1.

A Panther A or G in front of the Arc de Triomphe in Paris in 1944. Even without the French landmark in the background it would have been obvious from the anti-aircraft machine-gun on the turret that this photograph was taken in the West.

Right: A Panther D (recognizable by the cylindrical cupola and the loading hatch on the left side of the turret).

121

Panzer V - Panther

Sd.-Kfz. 171

FOREWORD

The Panther was, in terms of design and technology, probably the best medium tank of the Second World War, though it lacked only the maturity later brought about by long development and testing. Therefore it was not until 1945, when it was already too late, that the Panther began to show its true combat potential.

Spurred by the requests of our readers, we believe that we are responding to their wishes in producing this volume on one of the most famous German tanks.

Panthers assembled in a village behind the invasion front (Normandy) in 1944. This photo is interesting for several reasons, first, because it illustrates an *Ausf.* D (foreground) and, second, because the tanks carry individual names (on the folded-down gun travel locks). The *Ausführung* D could be distinguished from the A and G models by its lack of a bow machine-gun. This tank has the cast commander's cupola with its seven periscopes, which was rarely seen on the D.

THE PANTHER

New information has come to light, however on the 204 Panthers sent for the new tank's first — and from a technical standpoint premature — first action in July 1943 (Citadel), where one-fourth broke down during the drive from the railheads (Orel and Kharkov) to the operational areas (mostly engine fires and running gear damage). Part of the problem was also caused by the fact that constantly changing requirements had resulted in the Panther weighing 10 tons more than originally planned.

The first series, production of which began in December 1942, was designated *Ausführung* (Model) D. The second series, production of which began in July 1943, was designated *Ausführung* A. Through assimilating features of the Tiger and as a result of further experience, the *Ausführung* G appeared in 1944. In contrast to the D Model, the last two versions acquitted themselves well. Naturally, their success was aided by the better familiarity of the crews with their weapon.

The Panther was intended as the standard tank for regular Panzer Divisions. However, on production grounds only one battalion of each regiment was so equipped. The other battalions were equipped with the Panzer IV or *Sturmgeschütze* (assault guns).

In the course of efforts to simplify production, even greater commonality with the Tiger II (*Königstiger*) was planned from 1944. The new version was to be called the Panther II. The project got no further than the drawing board, however. Looking back, one can draw the conclusion that instead of producing four types of tank (Panther I and II, Tiger I and II), the production of a single medium tank, the Panther, would have sufficed. This was all that the units needed — despite the rather legendary reputation of the Tiger. So it is in the armies of today, which have only one type of main battle tank.

As was the case with the Tiger, examples of the Panther were produced as command tanks with increased radio equipment, a tank-destroyer version with a fixed superstructure was built and, finally, a *Bergepanther* (Recovery Panther). In total barely 7,000 Panthers were built (somewhat more than 6,000 as battle tanks).

Sd.Kfz.	Typ	Hersteller	Baujahr	Stückzahl
171	PzKampfwagen "Panther" (7,5 cm) Ausführ.D	MAN	1942-1943	ca. 50
267/268	PzBefehlswagen "Panther" (7,5 cm) Ausf.D.	MAN	1943	
	PzBeobachtungs-wagen "Panther" Ausf.D			Umbau (wenige)
171	PzKampfwagen "Panther" (7,5 cm) Ausf.A	MAN (ca. 1950)	1943-1944	
267/268	PzBefehlswagen "Panther" (7,5 cm) Ausf.A	Henschel (200) 1943	1943-1944	ca. 6.000 (1943-1945, all four)
171	PzKampfwagen "Panther" (7,5 cm) Ausf.G	MNH(1750)	1944-1945	
267/268	PzBefehlswagen "Panther" (7,5 cm) Ausf.G	Daimler Benz (ca. 1600)	1944-1945	
173	8,8 cm Panzer-jäger "Jagd-panther	Miag (270), MNH (112)	1944-1945	382
179	Panzer-Berge-wagen "Berge-panther"	Demag (Benrath)	1944-1945	297

Left and on the facing page are *Ausf.* D Panthers with the later (cast) cupola. All models of the Panther had the same 75mm gun (KwK 42 L/70).

In the upper right corner of the facing page is the division insignia of the Panzergrenadier-Division ***Grossdeutschland.*** This division was made up entirely of volunteers and was one of the first to receive the Panther.

Facing page: A Panther A leaving or moving onto a flatcar over an improvised ramp.

The 1st Panzer Division (division insignia in upper, right corner of the photo) received its first 115 Panthers in mid-1943.

Above: A Panther *Ausf.* A on a special rail car of the *Reichsbahn* (State railway). This photo is interesting as it shows three Panthers about to leave their rail cars at the same time over a wide end ramp.

Right: Panthers (probably also A Models) aboard a train. The tanks wear a multi-color camouflage finish. Interesting are the different exhaust arrangements on the two Panthers in the foreground.

Above right: The unofficial insignia of the 1st Panzer Division. It was often worn next to the official insignia (see facing page).

50

Left above: A Panther transport train. It is noteworthy that the tank's crew travelled with their Panther, pitching their tents on the flatcar beside the tank, and that the unit's wheeled vehicles were loaded aboard the same train.

Below left: A Panther in winter camouflage finish. As on most other types of Army tank, the size, shape and color of the Panther's identifying numerals and letters were far from uniform. There were black, white and red figures, large and small, with and without white borders. The codes always followed the same pattern, however. The tank illustrated below, for example, Number 232, belonged to the 2nd Company, 3rd Platoon, and was the 2nd vehicle of the platoon.

Facing page, right: Ammunition is loaded through the rear loading hatch (on the D Model there was also a loading hatch located on the left side of the turret). The composition of the ammunition load (armour-piercing and high-explosive shells) corresponded to the type of opposition expected, with armour-piercing always predominating.

Panthers (left, an *Ausf.* D; below, an *Ausf.* A) on the march. All versions of the Panther had a maximum speed of 46 kph and a fuel capacity of 730 liters in five tanks. Consumption was 450 liters per 100 km on-road, and 670 liters per 100 km off-road, giving it a range of about 160 km over roads and 100 km cross-country. By the standards of today's tanks the Panther's range was very limited.

Of these photographs of Panthers on the march, the one on the right is especially interesting. At the head of the column is a *Funklenkpanzer* (Radio-Controlled Tank) B IV (Sd.Kfz 301) which carried an explosive charge on the front of the vehicle. The 450-kg, prism-shaped explosive charge could be released by radio command. Visible on the Panther behind are two "log carpets" (sections of logs strung together, which could be placed beneath the tank's tracks to improve traction) for use in difficult terrain. It carries one on the glacis and another on the turret roof.

An *Ausführung* G Panther with the camouflage scheme commonly used in 1944/45. On a dark yellow base are patches of red-brown and green which are broken up by spots of the dark yellow base color.

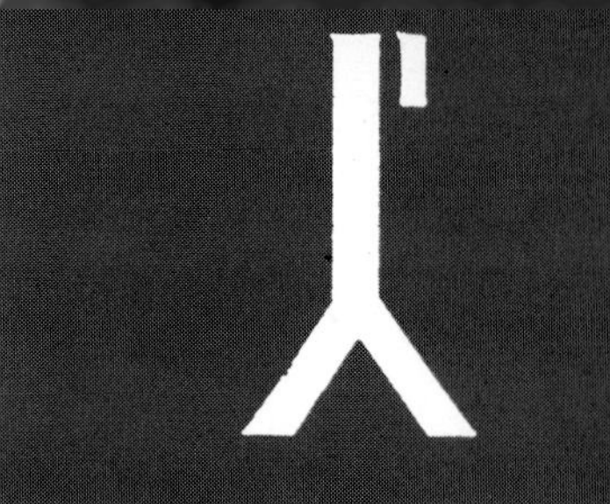

Right: The division insignia of the 2nd Panzer Division which was in use from 1941. From 1943 the Army's regular Panzer Divisions were issued Panthers; production allowed only one battalion per division, however, or about 100 tanks. The division's First Battalion was usually the one equipped with Panthers.

The front of a Panther A. The small opening for the telescopic sight is visible in the ball mount for the bow machine-gun. To the right of the closed driver's vision hatch is the insignia of the 23rd Panzer Division.

The Panther of the commander of a regiment, recognizable by the code 01. He is *Oberst* Langkeit (wearing the peaked cap), commanding officer of the ***Grossdeutschland*** Panzer Regiment. This tank provides a good illustration of the external stowage of various pieces of equipment (tow cable, cable cutters, crank, track-connecting tool, etc.).

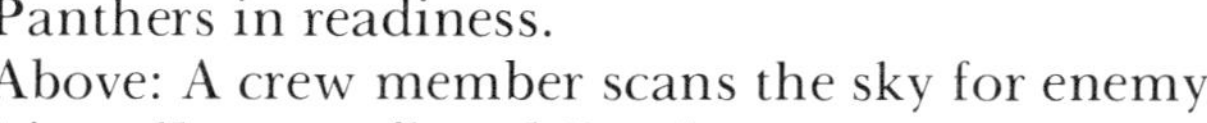

Panthers in readiness.
Above: A crew member scans the sky for enemy aircraft from his well-camouflaged Panther.

Above right: On guard in a village; in the foreground are fresh graves of the fallen.

Panthers in Italy leave their readiness positions.

PANTHERS IN ACTION

The photographs on this page show the Panther in action with other weapons, specifically in cooperation with Grenadiers, a style of fighting more common from 1943.

These pages again show Panthers — mostly A Models — in action, above probably at the driving school in Cottbus, where the replacement units of the ***Grossdeutschland*** Division were stationed. Panthers very rarely carried the *Balkankreuz* as a recognition marking, especially after 1943, when the emblem was usually omitted for reasons of simplicity.

Facing page: Shot damage to a Panther G: a demonstration of the effectiveness of its armor. On the same page is the later division insignia of the 3rd Panzer Division.

Below: A Panther G — without a bow machine gun (a rarity!) — in Italy carrying a "log carpet" on its turret and, right, Panthers of the same model in Hungary.

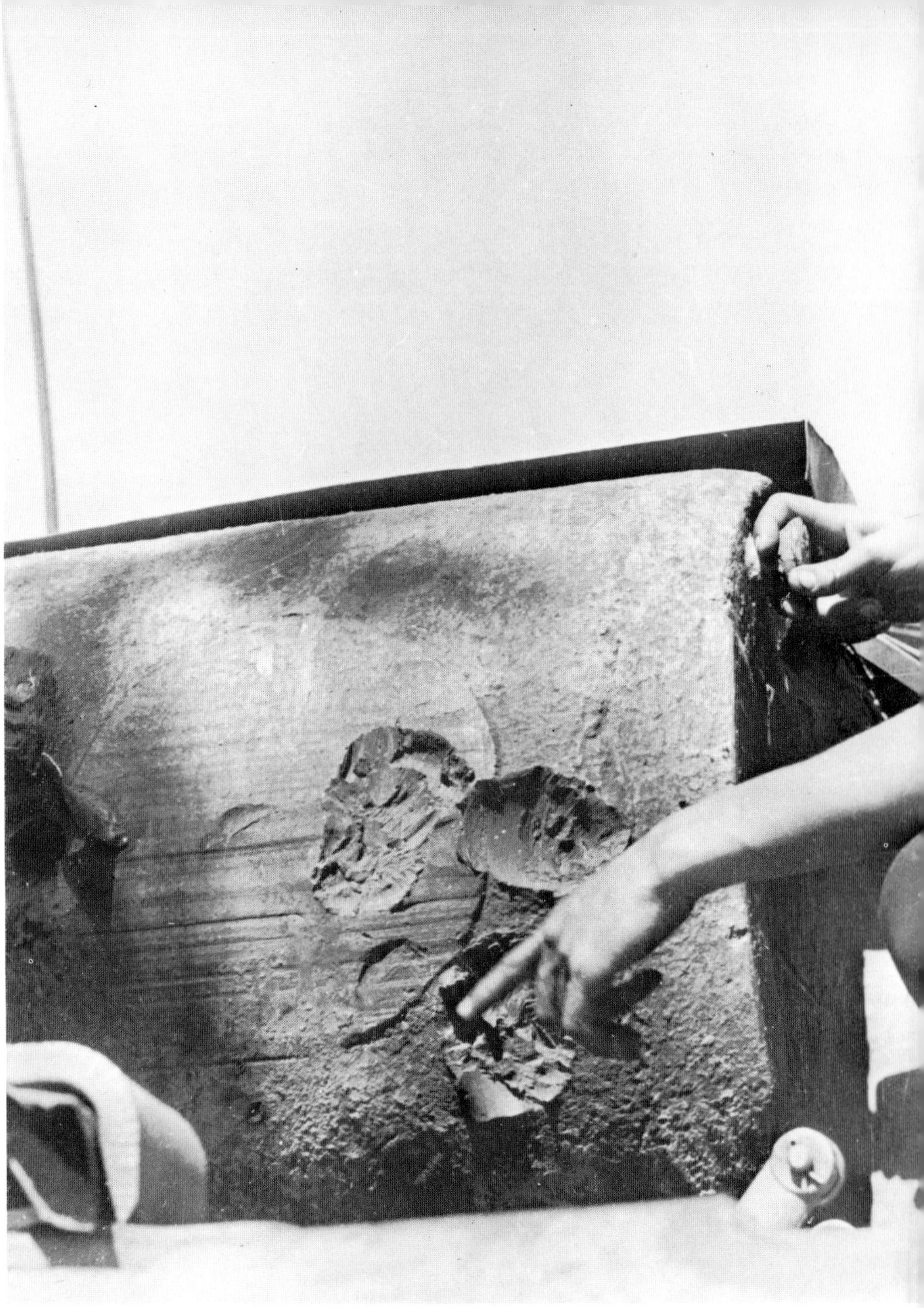

An *Ausf.* A Panther. The insignia on the left upper sleeves of the soldiers suggest that it belonged to an SS Panzer Regiment. An interesting note is the diversity of uniforms worn by the crewmen.

Below: Two Panthers in action, supported by infantry.

An *Ausführung* A Panther. It stands today in the outdoor museum of *Kampftruppenschule* of the Bundeswehr in Munster.

A Panther in action in Lithuania, summer 1944.

Left: A Panther — probably an *Ausf.* A — rolls around the site of the Arc de Triomphe in 1944. In the period just before the expected invasion German tank units often drove through Paris as a deterrent to local uprisings. The loading/escape hatch on the back of the turret is open and the AA machine-gun is in the ready position. Although the arm of the machine-gun mount proved to be rather weak, the installation, which was mounted on the circular ring on the commander's cupola, was effective. The commander has his headset on so that he can contact his platoon leader at once or, by turning the switch on his chest to "intercom," speak with any member of his crew (usually only with the gunner or driver).

Below is the division insignia of the 4th Panzer Division. All Panzer Division unit insignia were yellow on a dark background.

The photograph on the facing page shows well the width of a track and the close spacing of the individual links. This arrangement was adopted to reduce ground pressure and minimize the amount of dirt passing through the track to the running gear.

Facing page, far right: Detail views of the Panther's running gear.

155

PANTHERS IN WINTER

Left: This crew is wearing the padded, reversible winter uniform, which was white on one side and field-gray on the other.

Below left: Panthers take on supplies. Supply trucks have brought the necessary goods. Before the commander lies a folding map-mounting board; two clips hold the map, while next to it are a flare pistol and the commander's binoculars.

Below: The gun barrel is cleaned of powder residue.

Facing page: Victor and vanquished after a battle on the wintry steppe of southern Russia.

Left: The tanks assemble after the battle and await supplies.

Below left: In the mud of the early year.

Below: A Panther wearing makeshift winter camouflage.

Photographs from the Ardenne Offensive of December 1944. At this time nearly every Panzer Division had a battalion of Panthers.

Below right: A temporary bridge has failed to bear the weight of this Panther.

Left and above: Two further photos of Panthers being transported by rail.

An action photo from Lithuania. The sheet metal panels hung on the tank's sides, called *Schürzen* (aprons), were a protection against hollow-charge shells, which exploded against the panels and expended their energy in the space between them and the tank's hull.

An *Ausführung* D Panther with the cylindrical commander's cupola of the initial production batch.

Facing page: Panthers driving onto the battlefield in single file. On the horizon can be seen a platoon which has already deployed into wedge formation. The loading/escape hatch on the back of the turret is open for better ventilation. On the rear of all of the tanks are the inevitable pails for the engine and crew. As the commanders are standing in their turrets, it appears that enemy fire is not too heavy.

In the upper right corner of the page is the division insignia of the ***Panzer-Lehrdivision***. This division was formed in 1943 from instructional units of the *Panzertruppenschule*.

Above: A Panther camouflaged beneath trees. In addition to track links, it carries a spare roadwheel, pointing to one of the weaknesses of the Panther, namely the rim bolts on the wheel disks — they broke easily.

Right: A Panther in Italy. Its frontal armour was 8 centimeters thick, the remainder 45mm. Average speed over good roads was 35-40 kph.

L

Left: A Panther drives through a village in East Prussia. The turret number indicates the 4th Company, 1st Platoon, 4th Vehicle (of the platoon).
Above: Changing track links and the toothed rim of a drive sprocket. Both were subject to great wear as they had to stand up to tremendous forces.

Above: Panthers pass by Grenadiers.

A Panther with mounted infantry. This was rather dangerous for the infantry as the tanks tended to draw fire.

Above right: *Oberst* Stahl (6. Pz.Div.) in discussion with *Major* Pössl (Commander of the ***Grossdeutschland*** Panzer Battalion). The Panther shows the scars of a recent battle.

PANTHER Ausführung A

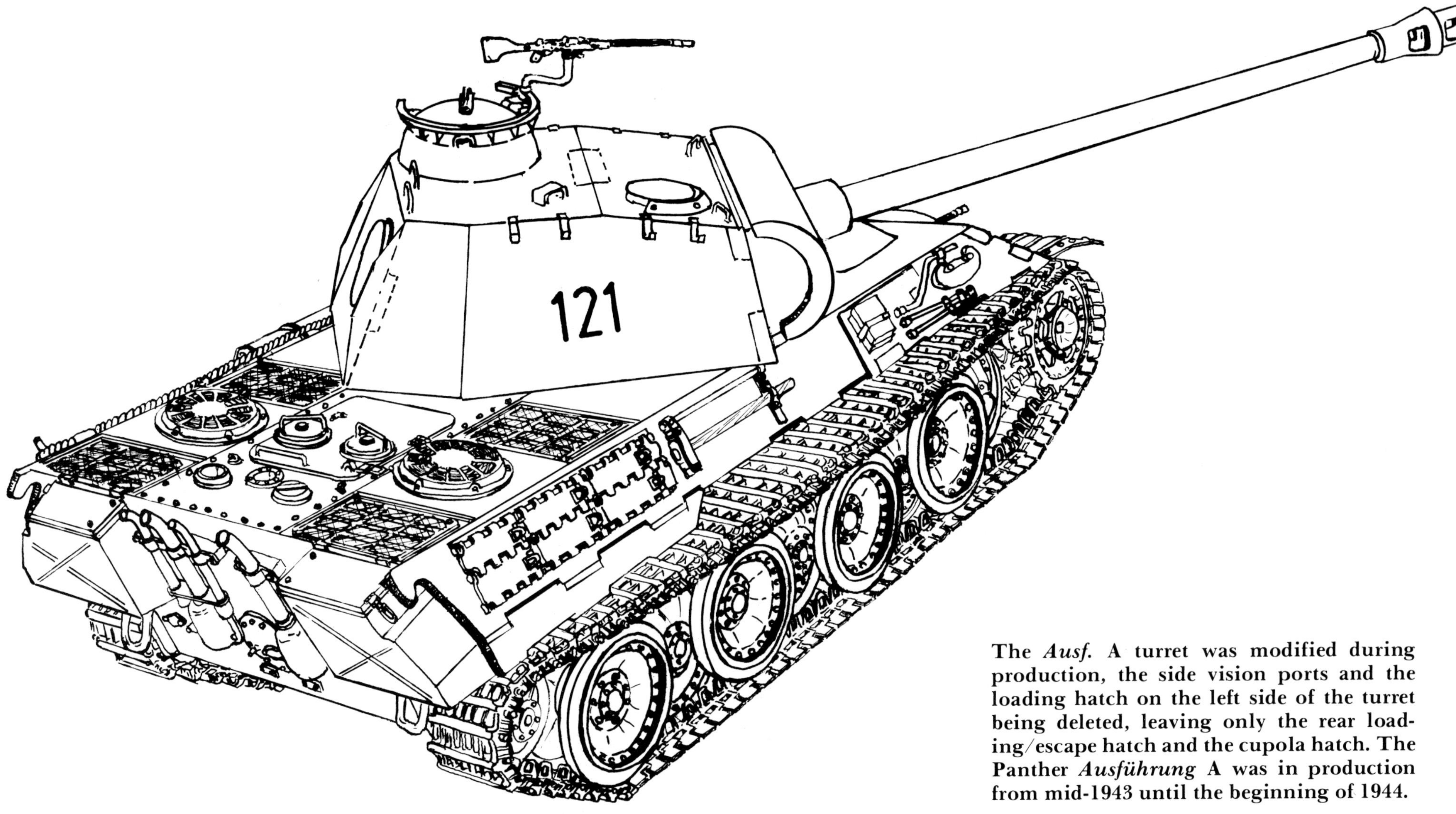

The *Ausf.* A turret was modified during production, the side vision ports and the loading hatch on the left side of the turret being deleted, leaving only the rear loading/escape hatch and the cupola hatch. The Panther *Ausführung* A was in production from mid-1943 until the beginning of 1944.

Panthers: above on the south Russian steppe, above right in a city in Courland and below on the invasion front in northern France. The latter tank is very well camouflaged on account of the massive Allied air superiority.

534

501

Both pages show Panthers in action — the photographs speak for themselves: an efficient external shape combined with powerful armour, 700 h.p. and a five-man crew behind an outstanding 75mm gun (KwK 42 L/70) and two or three machine guns. If the war had lasted long enough for the Panther to reach maturity it would undoubtedly have become the best mass-produced tank of the Second World War.

Left: A Panther of a Waffen-SS division on the steppe. A platoon leader marshals his tanks.

Below left is the emblem of the 5th Panzer Division, which was adopted in 1941. This symbol appeared on all of the division's vehicles, commanders' pennants and divisional road signs.

Below: A Panther watches over and supports an attack by dismounted Grenadiers.

Above: Discussion before an operation.

Above right: Insignia of the ***Hermann Göring*** Panzer Division. The division was formed from guard battalions of the Luftwaffe.

Panthers utilize a railway embankment during the advance on Wilna (1944).

R01

12

The end: knocked out, blown up, dug in (following irreparable engine damage), burning and finally as booty for the other side (facing page).

Of the 7,000 Panthers built only a few examples remain in tank museums.

• *Schiffer Military History* •

Specializing in the German Military of World War II

Also Available:

• The 1st SS Panzer Division - *Leibstandarte* • The 12th SS Panzer Division - *HJ* • The Panzerkorps *Grossdeutschland* •
• The Heavy Flak Guns 1933-1945 • German Motorcycles in World War II • Hetzer • V2 • Me 163 "Komet" •
• Me 262 • German Aircraft Carrier *Graf Zeppelin* • The Waffen-SS - A Pictorial History • Maus • Arado Ar 234 •
• The Tiger Family • The Panther Family • German Airships • Do 335 •
• German Uniforms of the 20th Century - Vol.1 The Panzer Uniforms, Vol. 2 The Uniforms of the Infantry •